OUR EXPANDING
UNIVERSE

THE MCGRAW-HILL HORIZONS OF SCIENCE SERIES

The Gene Civilization, François Gros.

Life in the Universe, Jean Heidmann.

Our Changing Climate, Robert Kandel.

Earthquake Prediction, Haroun Tazieff.

The Future of the Sun, Jean-Claude Pecker.

How the Brain Evolved, Alain Prochiantz.

The Power of Mathematics, Moshé Flato.

EVRY

SCHATZMAN
OUR EXPANDING
UNIVERSE

McGraw-Hill, Inc.

New York St. Louis San Francisco Auckland Bogotá
Caracas Hamburg Lisbon London Madrid
Mexico Milan Montreal New Delhi Paris
San Juan São Paulo Singapore
Sydney Tokyo Toronto

English Language Edition

Translated by Isabel A. Leonard
in collaboration with
The Language Service, Inc.
Poughkeepsie, New York

Typography by AB Typesetting
Poughkeepsie, New York

Library of Congress Cataloging-in-Publication Data
Schatzman, Evry L.
  [*L'Expansion de l'univers*. English]
  Our expanding Universe/Evry Schatzman.
      p. cm. — (The McGraw-Hill *HORIZONS OF SCIENCE* series)
  Translation of: *L'Expansion de l'univers*.
  Includes bibliographical references.
  ISBN 0-07-055174-X
  1. Expanding Universe. I. Title. II. Series.
QB991.E94S3213 1992
523.1'8 —dc20   91-30312

The original French language edition of this book
was published as *L'Expansion de l'univers*, copyright © 1989,
Hachette, Paris, France.
Questions de science series
Series editor, Dominique Lecourt

TABLE OF CONTENTS

Introduction by Dominique Lecourt   7

I. Our expanding Universe 23
    The paradox of expansion 23
    The galaxies 24
    The receding galaxies 27
    Expansion and relativity theory 31
    Quasars 37
    The background noise of the sky 38
    The theory of expansion 43
    Other models? 47

II. Open questions 51
    The question of isotropy 51
    The chemical composition of the Universe 53
    The inflationary model 56
    The unification of the forces of nature 57
    Broken symmetry 62
    From the cosmos to life? 63
    Distribution and formation of galaxies 66
    How old is the Universe? 71

III. Instruments and reflection 75
    Telescopes, computers, satellites 75
    The unity of the Universe 81
    Basic research 84
    The return of God? 93

Bibliography 99

INTRODUCTION

Within a period of only a few years, at the beginning of this century, our image of the Universe was turned completely upside down. An intellectual revolution occurred comparable in scope to those at the end of the 16th century and the beginning of the 17th century illuminated by the names of Copernicus, Kepler, and Galileo.

Certainly this upheaval was linked to improvements in observation, detection, and calculation equipment: large telescopes, radio astronomy, space exploration, and powerful computers are the most spectacular and best known examples. The role of instruments in the progress of knowledge is somewhat reminiscent, albeit on a totally different scale, of the role Galileo's telescope may have played in the triumph of the Copernican concept of the Universe. By developing the famous *perspicillum*, minutely described in his *Sidereus Nuncius* [The Sidereal Messenger] in 1610, Galileo opened the way to the discovery of mountains on the Moon, unknown "planets" in the sky, an immense number of fixed stars, in short: "...things no human eye had ever seen nor any human mind conceived of." Copernicus was correct: the stars are indeed composed of the same substance as terrestrial objects; far from being motionless in the center of the world, the Earth is only one of many planets.

Without the 2.5-m [100-in.] Mount Wilson telescope, the decisive discovery by Edwin Powell Hubble in 1927 that the galaxies were flying away from one another would never have been possible; without the availability of radio telescopes, Arno A. Penzias and Robert W. Wilson would not have been able to detect the "background radiation of the sky" in 1965. The hypothesis of the "expanding Universe" could have been neither formulated nor confirmed.

It is, however, the linkage joining each of these two revolutions to a major rethinking of the concepts and theories of physics, that allows us to relate them without artifice. The astronomical revolution inaugurated in 1543 by the famous work of Copernicus, *De revolutionibus orbium coelestium* [On the Revolutions of Heavenly Bodies], could not have borne its full fruit without dismantling the Aristotelian physics that had hitherto dominated the scene. Without the new concept of space imposed nearly a century later by Galileo and Descartes following a hard struggle, the new "world system" could not have triumphed.

Today, a true conjunction has taken place between the observational data accumulated since the 1920s and developments in relativity and quantum theory which, at the same time, had changed the face of physics. "Astrophysics," the new discipline born of this conjunction, combines known physical laws to interpret the properties of the heavenly bodies described by the astronomers; it applies these laws to invented models to account for observed systems.

Evry Schatzman, one of the pioneers of astrophysics in France, clearly describes in the opening pages of this short book the conditions under which investigations of the infinitely large and the infinitely small came together, joined, and became one. From this complex history, full of unexpected turns and twists, he isolates and describes with precision the prospects for the future, while at the same time emphasizing the danger facing the world of ideas.

This danger is the paradoxical aspect of this research: because it assumes an expanding Universe, it seems to resuscitate the cosmological concerns of antiquity.

If expansion indeed is taking place, it must be possible to assign a beginning to this process, and if there was a beginning, how can we not assume that there will also be an end? The 18th century, following the lessons of Newton, believed that science would now be free of these questions dogged by theology and metaphysics. Witness Immanuel Kant in his *Critique of Pure Reason* (1781): had he not demonstrated that the ultimate questions of cosmology—Is the Universe finite in space? Has it an origin in time?—are insoluble because they are illusory? They have no purpose, he explained, because they derive from an illegitimate transfer of moral and religious concerns, which make sense in practice, to a realm of knowledge where they have none.

And yet here they are once more, at the cutting edge of positive research rooted in the most sophisticated physics! The big bang theory is surrounded by cosmological speculations: once again, the order, unity, and origin of the

Universe are presented as themes of scientific contro-
versy. There are plenty of astrophysicists, some of the
best, who even believe they should yield to cosmogonic
temptation and tell the story of the genesis of the Universe. Could it be that the astrophysical revolution has a
sense of humor—an actual "revolution" with the original
astronomical meaning of the metaphor? Does it mark a
return of thought to its point of departure back in the days
of ancient Greece?

The questions that have stymied it at least invite us
to examine from a new perspective the long history of
which it is the provisional outcome.

There is reason to believe that Pythagoras was the
first to advance a spherical theory of the Universe with the
Earth at its center; he may have been inspired by Oriental
philosophy. In any event, it was Plato, in the 4th century
B.C., who embodied the first cosmological synthesis
known to us, in his *Timaeus*, inviting us to give a mathematical accounting of the movement of heavenly bodies
within a Pythagorean type of cosmology. Eudoxus of
Cnidus, his contemporary and disciple, would be the first
to attempt to carry out this program. Going back to the
idea that the stars moved in concentric spheres around the
Earth, he succeeded in the mathematical feat of accounting not only for the regular movement of the stars but also
for the "wandering" movement of the planets by a subtle
combination of appropriate spheres.

It was this finite and mathematically ordered cosmos that Aristotle was to inherit. He added to the number

of spheres nested within each other (no less than fifty-six), and described them as being guided in their respective movements by the movement of the outermost concentric sphere of the heavens to which the fixed stars are attached and which is moved by the "immobile prime mover" of divine origin.

Ptolemy, in the 2nd century A.D., relying on the extraordinary development in observational astronomy which he conceived, reinforced this concept of the Universe. The statement of his "mathematical conception," which the Arabs would later call the *Almagest*, dominated astronomy for fourteen centuries! We may well wonder why this misconception met with such enduring success while, of course, observations of the sky increased in number and became ever more precise.

The question becomes even more acute when we recall that, back in ancient times, another theory had been formulated and defended in the 3rd century B.C. by Aristarchus of Samos, who believed the Earth was a planet like the others, which both rotated on its own axis and revolved about the Sun.

The memory of the condemnations of Copernicus in 1616 and of Galileo in 1632 is so alive that we often explain it by the spiritual and temporal ascendancy of the Catholic Church and, more generally, of the Christian religion. But we must beware of rewriting history with hindsight; not only can the rejection of Aristarchus' concept by the Greek thinkers obviously not be imputed to the church, but we can even say that it actually did not take sides in cosmological disputes and throw all its forces into the battle in favor of the Ptolemaic model until a time when this model was entering into a decline.

If we follow the analyses of historians of science such as Paul Tannery and Pierre Duhem, we see that this model persisted for other and far sounder reasons.

In a world where, until the time of Galileo, there were no instruments of observation, the model of the celestial spheres accounts very well for observations made with the naked eye. Whatever may have been said about it, the ancient and medieval concept of the Universe was not in the least driven by pure speculation; on the contrary, it was rooted in numerous, repeated observations. Before it imposed itself on the mind, the figure of the sphere imposed itself on the eye.

For what do we actually see when we look up into the sky, if not a "vault" sprinkled with points of light separated from us by insuperable distances and all revolving together around the Earth, and then, contrasting with the regular movement of these "fixed" stars, the "wandering" (although equally periodic) movements of the Sun, the Moon, and the planets? Indeed, how could we fail to gain the idea of a radical difference between the natures of these stars, their pure light, the well-ordered mathematics of their movements, and the opaque and changing realities of Earth?

The extraordinary mathematical precision of the Ptolemaic system, and its ability to account for most of what was seen by a very confident system of geometry, did the rest and imposed itself on the best minds with the force of the obvious, which did not begin to crumble until the end of the Middle Ages.

But there is another reason that helps to account for this extraordinary longevity: the Ptolemaic system agreed remarkably well with the only physics known for centuries in the West, the physics of Aristotle. We know that this Greek philosopher, who thought of movement within the general framework of change, divided it into two kinds: natural movement and violent movement. According to this concept, there are only three kinds of natural movement, upward, downward, and revolving around the Earth.

Physics, metaphysics, and cosmology joined hands: the circular movements were deemed to be perfect; they suited beings who themselves were perfect because they were of divine essence (heavenly bodies). On the other hand, in the "sublunar" world, the world of generation and corruption, the elements (air, water, earth, fire) combined to impart upward or downward movements to bodies. By virtue of their internal disposition linked to their constitution, heavy bodies tended to return to their "own place," downward, where they were destined to come to rest if not disturbed by some violent movement; light bodies came to rest by moving upward.

As long as this physics dominated thinking, the opposition between the celestial world and the terrestrial world remained the rule. In accordance with the data of perception, the ancient cosmos—closed, ordered, and hierarchized—went unthreatened. Emancipation did not really begin until the end of the 16th century, and it took yet another century for it to become established.

In a famous book, Alexandre Koyré described the path that led from Copernicus to Newton's major work, the *Philosophiae Naturalis Principia Mathematica* [Mathematical Principles of Natural Philosophy], as the transition "from the closed world to the infinite Universe." It was indeed around the philosophical question of infinity that the astronomical revolution which began in 1543 was played out.

Nicholas Copernicus set the Earth in motion, dislodged it from the center of the Universe, and hurled it into the skies. He undermined the foundations of the traditional cosmic order: the human species was no longer at the center of the world, and the cosmos was no longer organized around us; we had lost our most reliable bearings. Such daring thoughts, such moral courage under the conditions of his time, are enough to assure him a glorious place in history. It is an insult to his genius to have presented his work as a mere mathematical model with no pretense at truth, as his publisher, Andreas Osiander, did not hesitate to do. But it is no insult to remember some of the constraints which he was unable to escape.

For not only did the world of Copernicus remain subject to hierarchical thinking, ordered around two poles of perfection, the Sun and the sphere of the fixed stars, but it was still a finite world, even if not understood in quite the same way as by the ancients.

Epistemologists point out that this attachment to the sphere of fixed stars was not "logical" and that his system in no way demanded it. Perhaps so, but will we ever find a great thought that satisfies their ideal of complete consistency? In any event, the fact is that Copernicus never asserted that the world of the fixed stars was infinite; he only said that it was immense, meaning that it literally defied measurement.

Johannes Kepler, whom we may look upon as the true creator of modern astronomy, was the first to reject the solution of Aristotle, abandon the circle for the ellipse, speculate about the reason for the movements of the planets, and formulate the laws that would henceforth bear his name. He too rejected the idea of an infinite Universe. His *Astronomia Nova* (1609) is quite explicit on this point: "This thought," he wrote, "brings with it I know not what secret horror."

That this repulsion was inspired by religious motives is beyond a doubt: Kepler, born in 1571 in Wurtemberg, aspired to become a priest and was possessed of an ardent faith. Even more than Copernicus, whom he had read enthusiastically in his youth, he consecrated to the Sun a cult tinged with mysticism. He believed that he found the presence of the Trinity in the order of the world: the Sun corresponded to the Father, the sphere of fixed stars to the Son, and the space between, filled with the celestial aura, to the Holy Spirit!

But the idea of an infinite Universe seemed to him too to be refuted by actual observation. The way in which the fixed stars are distributed throughout space varies from one region to another. Now, Kepler argued, if the physical Universe were infinite and uniform, this distribution should be uniform as well. Galileo's discoveries, which he welcomed with joy, did not lead him to modify his position: he continued to maintain that our moving, heliocentric world is a singular world as opposed to the realm of the fixed stars.

Truth to tell, the resistance to the image of an infinite Universe was so strong that Galileo himself refused to take sides on this issue, although he rejected the idea of a center of the Universe, be it occupied by the Earth or by the Sun. In the famous *Dialogo dei due massimi sistemi del mondo* [Dialogue concerning the two chief world systems] (1632), he has Salviati, his interpreter, say: "Neither you nor anyone else has ever proved that the world is finite and has a shape, or on the contrary is infinite and interminate."

The thesis of the infiniteness of the world had made some headway, however, as shown by the very vigor with which it was rejected; soon it would triumph. Already Nicholas of Cusa, the last great philosopher of the Middle Ages, often erroneously viewed as a forerunner of Copernicus because he stated, without evidence, that the Earth moved, considered the Universe if not actually infinite (*infinitum*), at least "interminate" (*interminatum*). But the ideas of this man, who was a papal legate and cardinal, lay in oblivion for over a hundred years.

There was quite a stir, on the other hand, when Giordano Bruno, born in 1566, had the extraordinary audacity to shatter the celestial spheres and to claim positively, we may even say joyously, that the Universe was infinite. Arrested by the Inquisition in Venice, he paid for this audacity with seven years of imprisonment in Rome and was burned in the public square on February 17, 1600.

Bruno found his inspiration in the Greek atomistic theory of Democritus, Epicurus and Lucretius. He used their most telling arguments to criticize Aristotle: "What would happen if you put your hand through the surface of the heavens?" But he often thought of making the infinity of the Universe into a hymn to the glory of God. "The excellence of God is magnified and the greatness of his empire manifested. He is glorified not in one but in innumerable suns."

Giordano Bruno was neither an astronomer nor a physicist. Should we read him as a modern thinker, ahead of his time? We may doubt this when we read his arguments, based on a magical and vitalistic conception of the world: his planets are living beings that move according to the free impulsion of their desires! Be that as it may, the question he had reopened with the grand style of the purest passion would continue to echo throughout the first half of the 17th century. His ultimate fate certainly had something to do with the caution of those who thereafter held to the same thesis if on different bases.

This was the case of Descartes who, for reasons having to do with his mathematicized concept of space, was the first to argue the infinity of the world with logical rigor. Playing on words, he reserved the term "infinite" for God; when speaking of the world, he said only that it was "indefinite." A simple precaution inspired by the fear of sharing the fate of "the unfortunate Bruno" or that of Galileo? Sincere conviction? It matters little. For Descartes, the ancient opposition between the motionless world of the skies and the moving world of beings was banished for good. Descartes constructed a new cosmological concept on the basis of the new physics. One may, as Voltaire so skillfully did in his *Lettres Philosophiques*, mock the "whirlwind" to which it gave birth—because Descartes identified matter with extension—and indeed deride all of Cartesian physics which, as Gaston Bachelard put it, is only a "novelized physics." Nonetheless, it marks a date in the history of philosophy.

For the thesis of the infinite Universe to be soundly established, a theory of physics was needed that was superior to that of Descartes in supplying the foundation for its major idea, that of identification or homogenization of celestial and terrestrial phenomena. This, as everyone knows, was the work of Newton who, combining a corpuscular philosophy of matter and a Euclidean geometric concept of space, forged the image of the Universe which was to dominate the mind for two hundred years. Destruction of the cosmos and geometrization of space went hand in hand: the concrete and differentiated space of "places" of Aristotelian physics was henceforth replaced by the dimensional space, homogeneous and abstract, of Euclidean geometry, which is considered real, without matter thereby being reducible to it. Now, no limits can be set on Euclidean space. Therefore, the cosmos will no longer be regarded as a finite whole, hierarchically ordered, but as an open Universe unified by the identity of its laws and its fundamental elements.

The identity of terrestrial physics and celestial physics was thus established: the fundamental law of attraction links the smallest and the largest bodies—atoms and stars—of the infinite Universe. It is the same set of laws that governs all the movements in the Universe: the law of an apple that falls to the ground and the law of the planets that revolve around the Sun. Astronomy and physics are connected in the same submission to (Euclidean) geometry.

But—need we remind ourselves?—Newton did not know the nature of the force of attraction and refused to "forge hypotheses" about it. Conceived of as real, because it was called for by the mathematical structure of his system, it appeared in his writings as an "immaterial" force about which he could say nothing.

Jacques Merleau-Ponty stressed this apparent paradox: under these conditions, there is no Newtonian cosmology, properly speaking. And one can speak of a "decline" in the idea of the Universe the day it is conceived of as infinite: it loses the unity of an object of thought. Immanuel Kant, who had, in his *Theory of the Heavens* (1755), made his contribution to the understanding of the formation of the solar system, only thirty years later concluded that the Universe did not belong in the field of scientific investigation. And when Pierre-Simon Laplace spoke of a "world system" (1796), it was the solar system whose theory he established, in the full awareness that this system is only an infinitely small part of the Universe.

Nonetheless, he left behind a set of statements inspired by Newton, constituting what one might call a kind of implicit quasi-cosmology. Organized around the concept of absolute space and time, it took shape explicitly in the wake of an enormous misconception; the reservations of Newton as to the nature of the forces of attraction were forgotten and they were interpreted as physical forces or material properties. Quasi-cosmology thus occupied the heart of a philosophy of nature which reigned undivided until the beginning of this century: the doctrine of "mechanism."

It took the appearance of non-Euclidean geometries, the blistering criticism of mechanism by Ernst Mach, and the genius of Albert Einstein for the formidable questions Newton had deliberately left unanswered, and which his disciples had believed settled, to be discovered once more. How could one say that space and time were empty frameworks with no relationship whatever to the "contents" of the events taking place within them? How, without the aid of God, could one account for immediate action at a distance?

With Einstein, another story begins. The notion of the Universe was truly reborn in 1915 with the theory of general relativity. At the moment observation revealed it to be immensely more vast than had ever been conceived, the idea once more imposed itself that the Universe must be finite! Were all those efforts to arrive at the image of an infinite Universe merely a long detour? Should one return to a certain idea of the "cosmos" and to the ideas of order and perfection that attach to it? Evry Schatzman ends by putting the question forcefully; it is an issue of science that raises a formidable philosophical issue. The expanding Universe invites us to a new way of thinking, beyond the classical opposition of the finite and the infinite. The reader will have all the information he or she needs to take this daring step.

Dominique LECOURT

I

OUR EXPANDING

UNIVERSE

THE PARADOX OF EXPANSION

Probably because it has an impact on the imagination, the expression "expanding Universe" has rapidly assumed a place in the cultural heritage of humankind. Correctly so, we might add, because it goes back to one of the key notions of contemporary astrophysics. Twenty years ago, it was still treated with reservation by numerous experts, including some of the best. Today it is almost universally accepted: it is even called a "standard model."

It may be, however, that its success outside scientific circles is actually based on a profound misunderstanding: the "expansion" as conceived by astrophysicists defies common sense in many ways. In truth, the very word "expansion" is a misnomer, and if it had not been sanctioned by usage for sixty years or so it would have to be replaced by a more adequate term.

Here, common sense follows etymology: anything which "expands" must in effect "step outside of itself" and into some external environment. The image comes immediately to mind of a balloon that is blown up and expands in the room in which we happen to be. But how can we speak of an "external" milieu where expansion would take place, if we are talking about the Universe? Obviously, the Universe can only expand within its own limits.

For millennia, quite probably, people's minds have reeled at the thought of the infinite within the unity of a concept; but the scientific theory of the expanding Universe brings about an acute form of dizziness that is unprecedented because it is based not on mere speculation but on an impressive accumulation of observational data; the data are coordinated by theoretical designs that are now soundly established on the very foundations of contemporary physics.

It is only when we are aware of these data and understand these designs that it becomes possible to formulate a correct idea of "expansion" when applied to the Universe. The story to be told is a long but exciting one, part and parcel of the prodigious progress in astrophysics that, over the last twenty or thirty years, has completely altered its appearance.

THE GALAXIES

The kickoff came in the 18th century with the discovery by the great British astronomer, William Herschel, of astronomical objects that had a nebular appearance but whose nature was not definitely known. It was also not known how far these "nebulae" were from the Earth.

It was not until the beginning of the 20th century that these "island Universes," in the words of Herschel, were correctly located.

At the beginning of this century, thanks to new observational instruments, bright objects were discovered in certain of these nebulae that had to be identified with exploding stars. This phenomenon was not unknown of itself; it corresponded to very ancient descriptions already given by the Danish astronomer, Tycho Brahe (1546–1601), and by the German Johannes Kepler (1571–1630): descriptions of the first "supernovas." What was "novel" was that such explosions were observed in nebulae. If one knew the power of the explosion, one could deduce the distance of the star. And so it was calculated. This was when the revelation occurred, the complete overturning of previous ideas: when the distances were calculated, it became clear that these astronomical objects were not located in our galaxy! So they are not, as Sir James Jeans (1877–1946) still believed when he wrote his book on cosmogony, masses of gas located in our galaxy, each one of which would have been a kind of planetary protosystem; rather, they are clusters of extragalactic stars.

At the outset, the expression "extragalactic nebulae" was used to describe these star clusters, since observation showed that their composition was blurred or nebulous. Today what is more striking is their similarity to our own Milky Way, or "Via Lactea." We know that the Milky Way is a great system of stars; we also know that the extragalactic nebulae are also great systems of stars that are very remote. They are thus all designated by the generic and universally accepted term "galaxies," and the Milky Way is also called the Galaxy. Hence the galaxies appear as vast systems of billions or even hundreds of billions of stars.

So that we have some point of reference and an order of magnitude, I will note that, among the nearer galaxies, the two Magellanic Clouds are located approximately 150,000 light-years away and the Andromeda Galaxy is two million light-years away. Both are visible with the naked eye even though the light year, which is the distance traveled by light in one year at a velocity of 300,000 kilometers [186,000 miles] per second, is equal to approximately 10,000 billion kilometers [6,000 billion miles] or 63,520 times the distance from the Earth to the Sun. Observation of the sky reveals the presence of innumerable galaxies: about two thousand billion are visible today and, in all probability, the space telescope will multiply this number by five hundred.

Galaxies, the constituent parts of the Universe, are characterized by their mass and their luminosity, but also by other physical properties such as their rotational movement. It is known today that they have different configurations depending on their speed of rotation. Their shape is either a slightly flattened spheroid, when it is called elliptical, or it is very flat. Some of those which are flat have a particular morphology characterized by spiral arms: these are the spiral galaxies.

For a long time it was believed that the differences in shape between galaxies were due to dynamic properties, whereby the elliptical and spheroidal galaxies would rotate more slowly than the spiral ones. Another property remained unexplained, however: the former type contain scanty amounts of interstellar material, in contrast to the spirals.

Today we have quite a different explanation, one that is more surprising but more satisfactory. We know that the galaxies are not regularly dispersed in the Universe; instead they are clustered by the thousands into vast systems. Analysis of galactic movements within these systems has revealed a phenomenon that is extraordinary in its scope and, we might say, its gigantism: collisions between galaxies. The elliptical and spheroidal shapes are the product of such collisions during which the stars are redistributed in the galaxy resulting from this strange type of phagocytosis, while the interstellar material are expelled. Other events of the same type have been observed that support this interpretation.

Of course the very idea of a history of clusters of galaxies implies the notion of the age of the systems and provides benchmarks for the history of the Universe.

THE RECEDING GALAXIES

Until the early 1920s, astronomers were busy identifying these objects and establishing their nomenclature. The number of stars in them, and their distances, had to be determined. People were working on these inventories and refining research methods when a new revolution in thought occurred, just as decisive as the previous ones. Once again it came from observation. In 1927, the American astronomer Edwin Powell Hubble discovered that the more distant a galaxy, the greater its velocity relative to our own; moreover, he stated that there was a proportionality relationship between velocity and distance.

This discovery has often been recited. It is based on a property known to all wave movements that was established in 1842 for sound waves by the Austrian mathematician J. Christian Doppler, a professor in Prague: this property has come to be known as the Doppler effect or the Doppler-Fizeau effect. In fact this effect can easily be observed: the sound of any vehicle engine (an automobile or motorcycle, for example) seems higher in pitch when the vehicle is approaching and lower when it is driving away. In technical terms, the wave emitted by an approaching sound source seems to us to have a higher frequency than if it were motionless; conversely, when the source is receding, the frequency seems lower. This is readily understandable: in the first case, the time interval separating reception of two consecutive wave crests is decreased, since the second crest has to traverse a shorter distance than the first; in the second case, the opposite occurs.

Let me clarify this further by evoking an image I have already employed in this regard in *Les Enfants d'Uranie* [The children of Urania] (Seuil, 1986). Standing on a river bank, I throw in at regular intervals, for example every second, a dead leaf which will drift with the current. If the current is flowing at three feet per second, I am creating a line of dead leaves separated from each other by a distance of three feet.

Three hundred feet away, a motionless observer will see the leaves approaching and floating by at the rate of one leaf per second. If I now assume that this observer is walking upstream toward me, he or she would see the leaves go by more frequently than before. If the observer were walking at the speed of three feet per second, the observer's speed would be added to that of the current and the leaves, still three feet apart of course, would seem to go by at the rate of one leaf every half second; the apparent frequency increases when the current is flowing toward the observer.

If the observer now turns around and walks away from me, for example at a speed of one and a half feet per second, the leaves will this time drift by more slowly, at the rate of one leaf every two seconds as the speed of the observer is subtracted from that of the current. If the observer quickens his or her pace and walks down the river at the speed of the current, three feet per second, the observer will always be level with the same leaf. He will no longer see the leaves float by. The time interval between the passage of two leaves has become infinite. In the same way, you cannot talk to someone who is moving away from you at a speed greater than the speed of sound.

What is true of sound waves is also true of light waves, with the restriction that one cannot exceed the speed of light.

Observing the spectrum of the galaxies, Hubble noted that these galaxies were "shifting" toward the red:

they appeared to be redder than they were. We know that red light has a longer wavelength than the average wavelength of visible light. Thus we must conclude that the galaxies are moving away from us. Hubble then determined that the rate of this receding movement was proportional to the distance between us and this galaxy; he calculated a proportionality coefficient called the "Hubble constant."

In a simple geometric representation, that is, a Euclidean Universe, this amounts to saying that all mutual distances expand in the same way in the course of time. This at least is a phenomenon which can easily be represented: if you take a balloon and draw a grid on it and then inflate the balloon, all the squares will dilate. It is obvious that the further apart from each other two points were to begin with, the more the distance between these two points would increase.

With this basic discovery of Hubble and the receding galaxies, the idea of the "expanding" Universe began to take hold, even if the process could not yet be seen to be universal and if, failing adequate observational tools, we could simply think that we were surrounded by a world of galaxies racing away from us. It was only when Sir Arthur Stanley Eddington (1882–1944) connected this experimental discovery with the models of the Universe that had been in existence since the early 1920s that we had to recognize expansion as a universal phenomenon. We will now see under what conditions this is so.

EXPANSION AND RELATIVITY THEORY

When we evoke the idea of the expanding Universe, we often associate it with the theory of general relativity, namely the relativistic theory of gravitation formulated by Albert Einstein in 1915. But, as we have just seen, the expansion process was discovered quite independently of the theory of relativity; it was only because Einstein's concept later played a very important role in the continuation of research and the interpretation of results in this field that a place had to be accorded to it which, historically, it did not occupy. As far back as 1895–1896, H. Seeliger and C. Neumann, independently of each other, had shown, purely theoretically, on a strictly Newtonian basis, the "logical" necessity of an expansion thesis. Both had studied what happens in an indefinite, homogeneous, uniform, and self-gravitating medium, namely, when the various parts attract each other according to the universal law of gravitation. They had dealt with the problem in the most classical fashion, referring to Newtonian and Galilean mechanics. Now, when we consider this very simplified and almost abstract framework, we see that this uniform media that extends to infinity can only be in a state either of collapse or of expansion; there is no third option. Seeliger and Neumann showed that this property is due to the fact that any particle in this Universe is attracted by any other particle according to Newton's law. It emerges from their work that there is no force capable of withstanding the power exerted by the various layers of particles on each other; they all weigh upon each other and, as they attract, they tend to exert forces that converge at each point.

I repeat: contrary to what has often been said and written and which is still maintained in popular science books, the phenomenon of an expanding, homogeneous, uniform, and isotropic Universe is not in the least a consequence of general relativity; it is a consequence of the law of gravitation and is already found in Newtonian models of the Universe.

Can we then say that the theory of general relativity played no role in this research? No, of course not. But we must be clear on this point and understand the conditions under which this theory was applied to certain models of the Universe in the early 1920s. It was not straightforward, and proceeded by trial and error on the part of Einstein himself.

The idea advanced by Einstein was, if one dare say so, very simple and in essence very much inspired by the mechanical theories of the Austrian Ernst Mach (1838–1916). Einstein stated that the established laws in this realm are not independent of the matter which they affect; they are a product of this matter itself. Now, if this is the case, a constant such as universal gravitation must be subjected to a kind of causality coming from the entire set of masses of matter existing in the Universe; Einstein regretted that he never achieved this goal with the general theory of relativity.

But in order to confer a kind of comprehensive consistency on his system and avoid having to introduce initial conditions or boundary conditions into it, Einstein believed it was essential to add to his equation a "gravitational repulsion" term or a "cosmological constant": thus he obtained an indefinite steady-state system in which the effects of repulsion and attraction offset each other exactly. This was how his model looked in 1917.

The first to draw attention to the difficulties associated with the "cosmological constant" of Einstein's steady-state model was the Belgian priest Georges Henri Lemaître. He showed that this model was not as stable as its author would have liked, and that it would take only a gentle push for the steady-state Universe in question to start to contract or expand. Thus Lemaître constructed a cosmology based on the idea of a "primeval atom"—a Universe that had a very small radius a very long time ago. After a long period during which it was close to the Einsteinian state, where repulsion and attraction offset each other almost exactly, the Universe is believed to have gone into an expansion phase which would be the phase we know today. The idea of expansion was established on the very foundations of a system designed to exclude it!

It must be added that the Russian geophysicist Alexander Friedmann (1888–1925) had shown at about the same time that Einstein's equations could lead to a whole class of solutions which were not static, given the hypothesis of a universal distribution of matter in the Universe. The unit of length would contract or expand according to a certain function R of a variable t.

This is where Eddington stepped in and the general theory of relativity actually made its entrance into the research field with which we are concerned. Eddington established a link between the work of the gravitation theoreticians (Lemaître and Friedmann) and the observational results obtained by Hubble. At this point, Einstein's "cosmological constant" appeared to be a useless and deceptive artifice, as its author readily admitted.

Based on this link, and this correction, an entire research strategy was then designed and carried out. It consisted first in trying to observe and study the galaxies at increasing distances. It was with this in mind that the 2.5-m [100-in.] Mount Wilson telescope was built in the early 1920s, allowing Hubble to prove the expansion of the Universe. The same strategy and the same thinking were responsible for the construction of a 5-m [200-in.] telescope that was completed and placed in service in the 1950s.

Of course, the point was not only to observe remote galaxies: their distances also had to be determined. To this end, a system for measuring the distances of galaxies was put together from start to finish. Establishment of such a system is simple in principle, but its implementation does present some serious difficulties. As regards measurement, the first thing is to select a standard reference object. This is relatively easy. But as we come to deal with increasingly great distances, the measuring system changes; hence it has to be connected to the next one and, by successive linkages, arrive at the farthest object.

A big surprise was awaiting astrophysicists when they had to conduct the connecting operations. Once the universality of the redshift of the galaxies was ascertained, and an attempt made to measure distances based on this shift, a striking paradox became apparent.

In Hubble's day, in the 1920s, the changes found in wavelength were tiny, amounting to a few percent at most. Then, just before World War II, spectral shifts of ten percent were beginning to be observed. Today, objects are observed whose shift involves a factor of three or four! Here is the paradox: as we have seen, the velocity of the object is proportional to the change in wavelength. If the rule of proportionality is applied, we find that, when the wavelength doubles, the object reaches the speed of light—which is absurd. If we return to the example of the floating leaves, the frequency is halved (and the wavelength doubled) when the velocity of movement away from the source reaches half the speed of the leaves. In the case of light, this means half the speed of light. A very long wavelength would mean that the source of radiation was moving away at a velocity close to the speed of light.

This time, the exact relationship introduces a property defined by Einstein's special theory of relativity, the theory he formulated in 1905 and which does not concern the geometry of the Universe. We know that this theory states an equivalence between mass and energy (according to the famous equation ($E = mc^2$). But the finite nature of the speed of light, which it establishes, has other consequences. When the velocity of the object comes close to the speed of light, the wavelength approaches infinity at the same time. In other words, the speed remains finite but the wavelength increases indefinitely. As "abstract" as this formula may appear, it must not be considered a mere mental exercise. It is a real property that can be observed today in particle accelerators: here, phenomena are produced which in fact change the time scale when we are dealing with a moving system in which the life-time of certain particles is multiplied by 2, 3, 4, or even 10 as their velocity approaches the speed of light. This pure construction of an extremely daring theoretical concept has thus become an experimental result that is reproduced thousands of times each day. It can even be said that the indefinite increase in the wavelength when the speed of the emitter approaches that of light is an absolutely essential property of high-speed movement. Thus, when we observe reddening corresponding to a wavelength no longer increased by ten percent but doubled, tripled, or quadrupled, this means that the speed of the source relative to the observer is approaching that of light. The relationship between the change in wavelength and speed no longer satisfies the elegant linearity Hubble had imagined sixty years ago.

In actual fact, the phenomenon proved to be far more complicated than he had believed; we are in a situation where, to give physical meaning to observational data, we must choose a model of the Universe. We often say that the interpretation is model-dependent. In fact, its meaning is not independent of the model created for it; there is no information that can be processed without at the same time asking questions about the system of representation adopted. We will come back to these difficulties in more detail.

QUASARS

Scanning the sky with radiotelescopes revealed radio sources with very small apparent diameters, smaller than the resolving power of the instruments used up to that point. Thus a new and unpredictable major step was taken in 1963 with the discovery of quasi-stellar objects, called "quasars" (a contraction of quasi-stars) and abbreviated as QSO (quasi-stellar object). The first QSO, discovered by A. Sandage, proved to be an object of a galactic type: a truly very special "galaxy" whose apparent velocity was close to 100,000 km [60,000 mi] per second.

The next step was to identify these radio sources optically.

The general shapes of these QSOs had been recognized. It became possible to set up a system of analysis to detect these objects by their optical appearance and not, as at the beginning, merely by the radio transmissions that could be received from them. Thus, today we have a list of slightly over three thousand quasars. This list is of course an open-ended one and, every year, new objects are added as the means of detection improve.

Among these quasars, we find quite extreme objects whose change in wavelength reaches a factor of five. A veritable feat of observation: in this way we manage to see spectral lines that normally appear in the far ultraviolet! Thus, our vision today can penetrate deep into the Universe: the greater the shift relative to visible wavelengths, the farther away the objects observed.

Quite logically, the question was asked whether, apart from QSOs, similarly very remote galaxies existed that resembled other, nearer galaxies familiar to us. The mere fact that we see "quasars" despite their enormous distances from us means that they are intrinsically highly luminous. Their luminosity is more than ten times that of normal galaxies. So, how are "ordinary" galaxies to be detected at such distances?

A highly original idea was proposed by Jacqueline Bergeron of the Astrophysics Institute of Paris. Noting that there are absorption lines in the QSO spectrum and knowing that there are absorbent gases in galaxies, she suggested that some of these spectral lines could be caused by gases in a galaxy that happens to be in the path of the light. Thus a search was made, in the field of observation close to a QSO, for one or more weak galaxies, and some very remote and very weak galaxies were found. But, unlike the QSOs, they are real galaxies. They are in the line of sight of the QSOs, which of course does not mean they are the same distance away. In any event, some are now known for which the wavelength has changed by nearly a factor of three, even though these galaxies are more difficult to detect than QSOs.

THE BACKGROUND NOISE OF THE SKY

The history of astrophysics contains a wealth of unexpected discoveries. Only two years after that of quasars came a discovery that is probably the most fascinating and the most exciting and provides a firm foundation for the hypothesis of the expanding Universe. It is the discovery of what is called the "background radiation of the sky."

The tale has been told many times since 1965. Indeed, it deserves attention, because it illustrates the role of chance in the history of scientific progress, at the same time again bringing to light the close cooperation in current astrophysics between observation and theoretical construction.

In 1964, two radio astronomers, Arno A. Penzias and Robert W. Wilson, were using the radio antenna of Bell Telephone Laboratories at Holmdel-Keyport, New Jersey. This antenna, built for other purposes, held promise for research: it had a 20-ft [6-m] reflector with very low background noise.

Penzias and Wilson tried to measure the radio noise emitted by our galaxy. It quickly became apparent that they were getting a little more noise than was to be expected in theory. At first, they thought that this excess could be due to the design of the antenna. To get around this difficulty, they started their observations at a relatively short wavelength, 7.35-cm. To their great surprise, in the spring of 1964 they detected the presence of a radio intensity that was quite definitely coming from the sky. At the same time they discovered that, after subtracting components varying as a function of time, season, day, and the passage of the galaxy through the sky, there remained a constant noise which came neither from the Earth's atmosphere, nor from the Sun, nor from the galaxy. It was called therefore "background noise of the sky."

Excitement in scientific circles peaked when the paper reporting this discovery appeared a little later. Very soon, many observations of this radiation began to be made at different wavelengths. It was established that it was thermal in nature, came from all directions equally (isotropic), and was similar to that found inside an enclosure, a "blackbody" whose walls were at the temperature of 2.72 K absolute (approximately –270°C or –454° F).

To understand the importance of this discovery, we have to return to the historical and scientific significance of the "blackbody." We must go back to the closing years of the 19th century, when physicists were trying to determine the equilibrium of radiation in an enclosure at a given temperature. In concrete terms, the question was formulated as follows: if you take a closed box whose walls are at a uniform temperature, what is the state of the radiation inside it? We know full well that when a body is heated, we perceive radiation; in front of a furnace, first of all we see the infrared which heats our face; then a reddening is perceptible in the throat of the furnace, and finally when the furnace is very hot it turns white.

The existence of radiation coming from a closed box was known; the point was to quantify it. This was a rather difficult and even somewhat paradoxical situation since, to observe radiation inside a "blackbody," you have to penetrate the enclosure, which would then no longer be closed. However, the situation is not an impossible one: if the opening made is extremely small by comparison to the dimensions of the enclosure, the disturbance of the radiation is infinitesimal and measurements can be valid and meaningful. Blackbody theory as it was developed at the end of the last century on the basis of classical physics led to an absurdity: the equations gave radiation of infinite intensity! Obviously, this was contradicted by experiment.

The difficulty was overcome when Max Planck (1858–1947) introduced a new parameter (since known as "Planck's constant"). The only way, said Planck, of reducing the theoretical absurdity to which classical blackbody theory led was to assume that radiation occurred in discrete quantities, which he called "quanta." Thus Planck resolved the contradiction between theory and experiment in expressing the properties of a blackbody.

It must be added that in 1900, when Planck advanced the idea of "quanta," he held on to the radiation theory in the form in which it existed and, if one may say so, confined himself to adding a supplementary constraint. It was only the work of Einstein that compelled a complete overhaul of radiation theory a few years later.

The background radiation of the sky, also called "fossil radiation," raised another question. It had been discovered; it had been measured, despite the difficulties due to its low intensity; it was known that it was blackbody radiation. Now the task was to find its source.

The answer came almost immediately from a completely different line of research, not observational but theoretical. Investigations had for a relatively long time focused on the abundance of elements in the Universe. An extremely daring idea had been advanced, particularly by an American of Russian origin, George Gamow, in 1946: the abundance of elements in the Universe must be related to a very primitive state of the latter—a state in which it would have been at the same time very dense and very hot, one from which, by expanding, it would gradually have become less dense and far cooler. At the time, Gamow predicted the presence of a vestige of this initial radiation at a temperature of 6 K. If we compare this prediction to the 3 K now observed, bearing in mind the scant amount of data then available, we have to admit that this precision was quite a tour de force.

The astrophysicists R. Dicke, D. Wilkinson, and P.J.E. Peebles, working at Princeton University, were aware of Gamow's hypothesis. They made the connection between the "background noise of the sky" and the theory that the Universe must have passed through a hot phase (approximately 10,000 K).

Gamow's theory provided the key to radiation at 3 K without the necessity of adding another hypothesis; radiation at 3 K confirmed the theory of the expanding Universe advanced by Hubble. Everyone knows the methodological principle formulated in the 14th century by the English logician William of Ockham: there must be no more hypotheses than there need to be. Here, from the epistemological standpoint, we have a perfectly "Ockhamian" rational coordination of two preexisting theories by a new observational fact.

Henceforth, "expansion" moved beyond the stage of hypothesis and took on the dimensions of a true scientific theory that formed the basis for new research and which, today, is no longer contested even if its meaning is still debated.

THE THEORY OF EXPANSION

Now that we have witnessed its intellectual genesis, and before we examine its development, we should try to understand this theory itself. It is an understanding that is not immediately self-evident and demands considerable effort, because it requires us to discard some very old thinking habits.

The first difficulty, and probably the most serious one, has to do with our habit of separating the two variables, time and space. When we do this, at every moment in time we picture an instant photograph, as it were, of the Universe seen in a slice of time that would be the same everywhere.

This concept may seem obvious, but in no way does it tally with the process of observation itself because, as we go back along a beam of light, we necessarily go back in time as well.

So, we must discard from the everyday picture of the Universe this disjunction between time and space that is constantly adopted by the observer unwittingly. We attach very little importance to the few thousandths of a second which separate Paris from a radio signal coming from New Zealand; this difference has hardly any effect on everyday life. But when we go back billions of years in time, the case is singularly different!

So, we must reason in two steps.

First of all, by adopting the usual way of thinking we follow in everyday life. As I said, we need to retain the concept of a photograph of the Universe at a given moment in time, so that at that moment it has the same density and the same temperature everywhere; we assume that on a very large scale (a billion light-years in our immediate environment) the medium is homogeneous and uniform in all directions and at all points. According to this concept, if we try to apprehend the evolution of the Universe, we say that this indefinite, homogeneous, and uniform Universe is in a state of expansion. This means that in the past, and everywhere at the same time, it was more dense. Since this radiation can be observed and measured, and is therefore accessible, we can also puzzle out what happened to it by going back into the past. We will then say that the Universe was more contracted, denser, and hotter, that there were more photons (more "light quanta"), and that the wavelength of the radiation was shorter. Going back in time in this way, we reach an epoch when the galaxies were touching each other. We might then think that a little further back from this moment, the galaxies did not yet exist as independent objects. We then extrapolate and say that the Universe was composed of a homogeneous and uniform gaseous medium, at a very early stage when it would have been identical to itself at all points.

Quite another form of reasoning must now be adopted. This time, let us consider emitted light and follow the light beam itself to understand what happens along its path. As we have already said, anyone who talks about displacement in space is at the same time talking about displacement in time. For the astronomer, there is an equivalence between time and space. Thus we can choose an arbitrary point of departure to serve as the basis for our reasoning: an era, for example, when the temperature was 10,000 K. So let us track a light beam emitted at that time. We realize that at the end of the path, traversed in an extremely short time, this radiation will have been completely absorbed; the beam will have disappeared. The reason for this is simple: it has to do with the properties of the medium in which the photons constituting the beam must move. Since every photon occurs at such a high temperature, each interacts too quickly with the gaseous medium, which is composed of free protons and electrons, to be able to continue along its path. So it is trapped in matter.

On the other hand, when we are in an era when the Universe is about 300,000 years old and the temperature is no more than 3,000 K, the photons emitted can continue to travel. The explanation is just as simple: as they advance, the state of the medium is changing because it is cooling down; thus protons and electrons can combine. The electron begins to revolve around the nucleus and a hydrogen atom is formed. Now, hydrogen atoms do not absorb photons of radiation at 3,000 K, while before that point free electrons had prevented them from moving any further along this path.

Hence the photon can continue on its travels and reach the observer. But by the time it arrives it will have traveled such a long distance that its wavelength is multiplied by 1,000 and, what amounts to the same thing, its temperature will have been divided by 1,000. We can then say that we are still receiving "blackbody" radiation; but while it may have been emitted at 3,000 K, the Universe evolved to such an extent while the light beam was traveling to Earth that, at the end of its journey, the temperature of the Universe was only 3 K. Thus, if we imagine light beams leaving the Earth in all directions, we finally reach a zone where the temperature is 3,000 K and photons have just begun to emerge: a veritable wall of light.

In this description, in terms of space-time, we arrive at a sphere surrounding us that resembles an original emission sphere, and this is what emits the radiation received at 3 K.

The most remarkable thing about this hypothesis is that, from beginning to end, it calls on only two kinds of astronomical data: expansion, and radiation measured at the present time. This is its strength; it requires the addition of no physics other than that used in the laboratory. By its very simplicity, it seems to assert that the background radiation of the sky be interpreted by a process of expansion of the Universe.

OTHER MODELS?

Understood in this way, does the idea of "expansion" inevitably set us on the slope of religious thinking? Does it contain, at least implicitly, the idea of creation, so that astrophysics would restore the rights and the powers of a god of some sort?

Remember that in 1951, Pope Pius XII, addressing the Papal Academy, declared that in the "primeval atom" theory of Abbé Lemaître he saw the original *fiat lux* ("Let there be light") and the proof of God's existence. Pius XII had visibly learned some lessons from the Galileo affair; he was anxious for the Church not to turn its back on the science of its day under the pretext of defending a literal interpretation of the Book of Genesis. Proposing a symbolic interpretation of the Scriptures, he could "recover" a certain interpretation of the idea of expansion for the benefit of religion. There has been no lack of Christian philosophers, such as Claude Tresmontant, who have followed this line of thought. Starting with the idea of Parmenides of a world always identical with itself, Tresmontant had to add to this visible, expanding Universe, which has a history, an additional part which ensures the eternal identity of the Universe and places God at the helm of the entire process.

This is probably one of the reasons why many astrophysicists who do not share these convictions and have misgivings about the intervention of religion into the field of their research have long been mistrustful of the very idea of an expanding Universe. As scientists, they have thus attempted to suggest alternative explanations. They have built other models.

This was the case, in the immediate postwar years, of the famous British astrophysicist Fred Hoyle, who, in the name of the "perfect cosmological principle" (which strongly resembles the principle of Parmenides), introduced the hypothesis of continuous creation. Within an expanding Universe, the appearance of protons and neutrons distributed randomly at the rate of one particle per 100 m^3 [about 3,500 ft^3] every several billion years keeps the density of matter constant. In one way or another (just how is not specified), new galaxies are born from this matter, but all in all the notion of origin, rejected in the name of flaunting atheism, is radically pushed aside for physical reasons.

Ingenious as it was, this hypothesis could not be proved by any existing theory of physics; to establish Hoyle's "continuous creation," an entirely novel "exotic" type of physics would have to be invented. The discovery of background radiation of the sky, as we shall see, was to deliver a fatal blow to this hypothesis. In 1970, Fred Hoyle was still attempting to rescue it, but without success.

This was also the case, twenty or so years later, of the Frenchman Jean-Pierre Vigier. Following him, Jean-Claude Pecker looked to hypothetical interactions between photons and the intergalactic environment for the cause that Vigier called the "aging of light," borrowing the term coined in 1929 by the Swiss astrophysicist

Fritz Zwicky. This would have explained the redshift of the galaxies with no need for any expansion hypothesis. But it must be admitted that so far Vigier's hypothesis has not agreed with observational data and, in particular, has not made it possible to explain the spectra of the remotest galaxies.

Let us add that a Universe without expansion—which would be the case if this reddening were explained by a phenomenon of microscopic physics—is a steady-state Universe. As shown by Neumann and Seeliger for a Newtonian Universe, and as the equations of the relativistic theory of gravitation prove, a steady-state Universe is unstable. It can even be shown, following the old argument of Jeans, that it is gravitationally unstable on any scale.

II

OPEN QUESTIONS

THE QUESTION OF ISOTROPY

Any advance in knowledge raises as many questions as it solves; this indeed is what is important. Much more important than the answers, always tentative, given to this or that problem are the undecided questions that attract the researcher's attention and ceaselessly claim his or her thoughts. This is why science is essentially progressive: each of its conquests calls for others.

The theory of the expanding Universe illustrates this point: there are swarms of open questions, and they are difficult ones because they call on an extremely sophisticated form of physics whose essential notions defy common sense.

When, under the circumstances we have related, Penzias and Wilson discovered the background radiation of the sky, the "radiation at 3 degrees," the point that got attention was that, coming from all directions, it exhibited the same properties in all directions. This is what is meant by the term "isotropic" (literally, in Greek, "the same wherever one turns"). It is by taking this remarkable characteristic into account that one can say that it corresponds to absolutely typical blackbody radiation at a temperature of about 3 K above absolute zero.

We have noted in passing that this isotropy is not absolutely perfect: it did not take long to discover that, in fact, the background radiation of the sky exhibits a very slight anisotropy, attributable to the absolute movement of the Earth relative to the Universe as a whole. At the speed of the Earth relative to the sky (30 km/s [19 mi/s] more or less, depending on the direction of observation), the anisotropy of the background radiation of the sky corresponds to an absolute velocity of the Sun of 300 km/s [190 mi/s]. This velocity produces an apparent radiation temperature difference of + 0.001 K in one direction and −0.001 K in the other.

In fact, this is only a simple "technical" difficulty which can be overcome by calculation and observation. And we can continue to say, with the small correction we have just described, that the radiation in question exhibits extraordinary isotropy.

But the explanation of this fact in itself raises a considerable difficulty, this time a theoretical one. How are we to explain that the physical processes that generated the background radiation of the sky, and which date back to the very early times we have defined above, can have a causal effect throughout the Universe, as the isotropy that has now been established seems to indicate?

The question is highly pertinent. At the time it was produced, the beam from the "sphere of influence" defining the causally linked domain represents—when brought down to the present era—a billion light-years. In other words, a billion light-years defines the present volume in which an interaction was possible, from the beginning of the expansion of the Universe until the time when radiation decoupled into gas. Now, the Universe appears to us to be homogeneous, on a scale ten times larger!

This question has not yet been answered, even though numerous hypotheses, as we will see, have been worked out by physicists; it is possible that these will eventually find a way to overcome the difficulty.

THE CHEMICAL COMPOSITION OF THE UNIVERSE

Analysis of observational data on the chemical composition of the stars in our galaxy and even of neighboring galaxies reveals a very striking fact: our galaxy is essentially made up of three-quarters hydrogen and one-quarter helium. All the other elements combined represent less than 2% of the material.

In this respect, we must note that the Earth is absolutely uncharacteristic in its reflection of the abundance of elements, because it possesses only a very small quantity of hydrogen, in the water of the oceans, and even less helium. This oddity is accounted for by a whole series of mechanisms by which elements were separated during the formation of the solar system and led to elimination on Earth of nearly all of the light gases, hydrogen and helium.

The important thing is that our galaxy has a chemical composition of remarkable uniformity. When analyses of the same type are made of other galaxies, the same chemical composition—the same abundance of helium—is found everywhere. This might seem surprising at first view!

If we once more consider the Universe at the time when it was a homogeneous and uniform medium, as we described it in the previous chapter, we know that when we go back in time we eventually find temperatures that are so high that nuclei can be transformed.

Around the time $t = 1$ s, when the temperature was on the order of 10 billion degrees Centigrade, hydrogen could rapidly be converted into helium and vice versa. The temperature affected the rate of the reaction. Now, the rate of helium formation varies slowly with temperature while the rate of helium destruction decreases very rapidly with temperature; at 10 billion degrees Centigrade, this maintained a chemical equilibrium between the elements. But rapid cooling during expansion stops the reaction between nuclei after less than a minute. In fact, the rate of helium destruction fell off so sharply that in a few seconds the abundance of helium froze. Thus, the chemical composition of the primeval Universe as we observe it was stopped short.

With this explanation, the uniformity of chemical composition of the Universe may be held to be one more argument in favor of the expansion theory. But once again a very formidable question looms, similar to that raised by the isotropy of the background radiation of the sky: the question of the causality that could explain this uniformity. Causality is used here in the most elementary sense of the term, because cause precedes effect. In macroscopic physics, the place and moment of the effect must be situated relative to the place and moment of the cause, so that a signal can propagate from one to the other at a velocity less than the speed of light. With such a definition, in an expanding Universe, the causally linked area is defined by a certain distance which is the speed of light multiplied by a time which is the "age" of the Universe in the era in question. Now, we can calculate fairly easily the mass of matter present in the causally linked area at the time the elements were formed. But we find that this mass adds up to no more than a few solar masses. This being the case, how can we account for the fact that an area as large as that of our galaxy—equivalent to 100 billion stars or several tens of thousands of times the mass of our galaxy if we include all the neighboring galaxies—can have the same helium and hydrogen composition, while its dimensions manifestly place it outside the causally linked area at the time the elements were formed?

This question of causality is, I repeat, a formidable one; it is probably the greatest difficulty that cosmology has to confront today. Of course, as with all questions of this type, researchers have come up with explanatory hypotheses. To be fair, it must be said that at this point we are treading on speculative ground. I will not deny that such speculation is exciting in every respect and may, at some unpredictable date, lead to the formulation of new experimentally verified concepts. But we have to say that today such a statement is still out of reach of contemporary astrophysicists.

THE INFLATIONARY MODEL

Here is one attempt at an explanation, presented schematically; it corresponds to what is now usually called the "inflationary" hypothesis. This puts forward the idea that in a far earlier era there was such a rapid expansion ("inflation") that the causally linked areas were far greater than what can be determined today. If we go down this path, we have to assume that the effects of causality that have been ascertained were determined well before the time when the light elements were formed, at a time when there would have been an enormous repulsive force that caused an extraordinarily rapid expansion beginning with a phenomenally high density. As we see, this reasoning is based on a hypothetical process believed to have been located "this side" of the known expansion, one which would hence have to be described in quite a different way.

This comment has a very precise meaning: it means not only that there is no experimental evidence for this process but also that, to support it, we would have to resort to truly "exotic" physics, quite different from the physics currently in use; on the other hand, if we content ourselves with going back to the origin of the elements, we have a sound, tested basis in physics, even if it is not of noteworthy simplicity. What is more, we know nothing about the primeval elements, assumed to have been ultrasensitive in nuclear terms, which we assume to have existed at that time.

This theory of inflation was invented by the physicists, and the astrophysicists have no observational data to contribute in its favor. Its authors are quite ready to admit that for the time being it is a kind of game. One might wonder what would be the point of physicists playing this game. In physics, the matter is a highly serious one; it is viewed from the angle of what is called the "unified field theory," namely the theory of the four fundamental interactions of nature.

THE UNIFICATION OF THE FORCES OF NATURE

Gravitational forces govern the movements of planets, stars, and galaxies while electromagnetic forces govern the structure of atoms and the production of light. But in nature there are two other forces, discovered by physicists when they investigated the atom: "strong" interactions and "weak" interactions. These four types of force are the fundamental forces of nature and are manifested throughout the Universe.

We should try to understand the strong and weak interactions as precisely as possible if we want to understand the physicist's goal of unifying these four forces within the framework of the "unified field theory," and at the same time understand the reasons for their value in astrophysics and cosmology. I will then describe how the approach of the astrophysicist cannot be confused with that of the physicist without grave misunderstandings.

57

It had already been known for some time that atoms are made up of negatively charged electrons balanced by an equal positive charge, when Lord Ernest Rutherford showed that the totality of this positive charge, also responsible for 99.95% of the mass, was gathered at the center of the atom, hence the name "nucleus."

As we know, it was in 1932 that Sir James Chadwick made the experimental discovery of a neutral particle (the "neutron") which, with the positively charged proton, formed the interior of the nucleus. Neutrons and protons were subsumed under the name of baryons.

But in the particular field of chemistry that governs the relationships between the components of the atomic nucleus, it was found that the mass of the nucleus is less than the sum of the masses of the protons and neutrons; an astonishing situation and, in any event, one very different from what we know to be the case in "ordinary" chemistry, where the mass of the molecule is equal to the sum of the masses of the atoms of which it is composed.

The equation $E = mc^2$ established by Einstein between mass and energy allows us to understand the significance of these differences in mass, as it clearly suggests that an energy is involved.

Attention accordingly turned to the energy binding the particles in the nucleus. If we take the current and convenient image of a building and compare the protons and neutrons to bricks held together by a particular force acting as mortar, we see that this force must be particularly powerful to prevent the positively charged protons, which repel each other, from flying apart. It must be a far greater force than the gravitation that holds the Earth in its orbit around the Sun, or the electrical charge which holds the electrons around the nucleus. It is this very efficient attraction which is called "strong interaction."

But with the discovery of radioactivity by Henri Becquerel (1896) and its study by Pierre and Marie Curie, physicists noticed that in one of the modes of nuclear transformation, electrons were emitted. For historical reasons, this transformation received the name of "beta radioactivity." When sufficiently precise measurements were made, and everything was added up, it was noted that the energy carried away by the electrons so emitted was less than the energy available in the nucleus. Two explanations were possible: either the law of conservation of energy did not apply inside the nucleus of the atom, or an invisible particle—or, at any rate, a particle very difficult to detect—was carrying away the missing energy.

Certainly, no one was ready to lightheartedly abandon the law of conservation of energy which gives physics its well-known, impressive overall consistency. Thus the hypothesis of a new particle won the day. Since it had no electric charge and apparently no mass, the Italian physicist Enrico Fermi gave it the name "neutrino."

The next step was to find it experimentally. This was accomplished by Cowan and Reines in a famous experiment conducted in 1956. The experiment was a real achievement because the difficulty of catching a neutrino "on the fly" is enormous: it interacts very little with the other particles and only at a very short distance.

Thus, a new interaction, different from the others, was discovered and labeled "weak interaction."

Today we know that the transformation of hydrogen into helium in the Sun is accompanied by the production of neutrinos; both of these processes prove that weak interactions play a role in the hearts of stars as well as in experiments on Earth. The flux of neutrinos from the Sun was measured in an experiment performed by Davis at the bottom of a gold mine in South Dakota. Davis' results have recently been confirmed by an experiment of a different kind, also conducted at the bottom of a mine, in Kamiokanda, Japan.

The ambition of physicists is to bring together in a single theory—the "unified field theory"—gravitation, electromagnetism, weak interactions, and strong interactions. Their efforts were crowned with remarkable success as far as electromagnetic forces and weak forces are concerned with the discovery in 1983 at the CERN [European Center for Nuclear Research] in Geneva of W and Z particles, with the theoretically predicted masses.

The prevailing opinion is that such unification probably involves non-nuclear particles with a very great mass. Now, the presence of these very heavy particles, even their temporary or fleeting presence, can be imagined if we go back to a remote era of our expanding Universe. With higher and higher temperatures, the appearance of strange particles of this type is quite likely. Hence, it is believed that in the case of the first moments of the Universe, arguments could be found in favor of the unified field theory. It is their need for consistency, a very legitimate quest for physicists in establishing the relationship between fundamental interactions, that leads them to turn to astrophysics. And we can confidently assume that, if such a unification were brought about by them, astrophysics would benefit greatly.

One essential idea is to imagine a very long-ago past, without particles, when the "vacuum" contained a density of extraordinary energy. This energy of the vacuum would have allowed ultra-fast "inflationary" expansion until expansion of the Universe converted this energy into particles. Immense linear formations, in the form of residues from this era—called "cosmic superstrings" because of their shape and possessing an enormous power of gravitational attraction—would have lasted long enough to initiate the formation of galaxies. Their presence at the time of background radiation of the sky would have produced fluctuations in the environment, hence in its emissivity. The fluctuations in the background radiation of the sky due to these heterogeneities can be calculated and, one hopes, compared with observational data. But we must improve the precision of measurement by a factor of 100 before this model can be put to the test of experimentation. If this becomes possible, it would be the first observational test of the "new" physics.

Bear in mind, however, that for this latter point of view we have no trace today of the processes whose existence physicists are led to assume in their grand intellectual game.

BROKEN SYMMETRY

Let us consider once again the background radiation of the sky at 3 K which is a major argument in favor of the expanding Universe. If we go back in time to an era when the density of radiation energy dominated the entire system, we see that the number of photons per unit volume is about the same as the number of all the other types of particles, give or take a few, which is easy to explain. In other words, per unit volume there are as many photons as electrons, neutrinos, hydrogen nuclei, and neutrons combined, accompanied by their antiparticles.

A characteristic feature of the present situation is that the ratio between baryons (neutrons and protons) and photons is entirely different: for every hydrogen nucleus there are between one and ten billion photons! And we find that the ratio of the number of photons to the number of baryons remains constant over time. So how did the destruction occur? How can we account for such asymmetry in the process? And, another striking asymmetry: how is it that in our galaxy and the neighboring galaxies we are dealing only with matter and never with antimatter? How is it that we are dealing with particles, never with antiparticles, which we know full well are produced daily in great accelerators and which "accompanied" particles in the primeval Universe? These questions are still waiting for an answer. Various hypotheses have been put forward to account for this broken symmetry which is suspected of corresponding to a quite fundamental physical process. Stephen Weinberg proposed one; Jean-Marie Souriau another. There are others who find this broken symmetry in the neutrinos; it was demonstrated in the laboratory a few years ago. While the electron, in fact, can rotate in both directions, like its antiparticle the "positron" or positive electron, the neutrino rotates in only one direction, as does the antineutrino.

Whether, with Weinberg, we imagine broken symmetry as a microscopic process simultaneously affecting the entire expanding Universe, whether we look at it with Souriau as a macroscopic phenomenon effecting a separation between matter and antimatter, or whether we scrutinize the strange behavior of the neutrinos, we may compute that this break could be at the origin of the tiny imbalance between matter and antimatter that had to exist in the primeval Universe if we are to account for its present existence.

As to the subsequent evolution, the problem is less complicated. Hubert Reeves in an elegant demonstration showed that baryon-induced entropy is relatively high, which means that the state of disorder of the Universe considered per unit mass is not very different from what it was at time $t = 100$ s, one hundred million years ago. I am in full agreement with this demonstration.

FROM THE COSMOS TO LIFE?

But the question of symmetry, or more precisely the question of broken symmetry, is worth dwelling on. The question we have just asked on the cosmic scale and which, as we have seen, is responsible for some of the most exciting research in microphysics, may be examined side by side with another question in a totally different field. The comparison seems to me to have a very exciting future.

Molecular biologists have found that, of the twenty amino acids existing in living beings, only one is symmetrical and nineteen can exist in a leftward or rightward form, mirror images of each other. It turns out that living beings are predominantly made of leftward or "levorotatory" amino acids, and in general of leftward molecules.

This strange predominance of levorotatory molecules has been the subject of research for a century now; Pasteur had been concerned with the problem as far back as the experiments performed in 1849–1850. He suggested in his notes that this asymmetry could be attributable to "cosmic asymmetry." This was of course only a general and venturesome view. But it is known today that, as we have just seen, there is a cosmic asymmetry in nature.

To make things clearer, we have to turn to the intangible, to the very subtle neutrino. The interaction between electron and neutrino is due to a very tiny force called the weak interaction. The neutrino rotates on its own axis, with the particular feature that its axis of rotation is always parallel to its velocity and the direction of rotation is always the same. If we consider its image in a mirror perpendicular to the direction of its movement, the direction of rotation is retained in the mirror image while the direction in the mirror is reversed: the mirror image of the neutrino is that of an antineutrino which is said to be turning rightward while the neutrino is turning leftward. Symmetry is broken.

Now, such a phenomenon affects electrons. Electrons, which are negative, have a slight tendency to turn leftward, with the direction of the axis of the spinning electron usually oriented opposite the direction of its speed, while the axis of the spinning positron (the antiparticle of the electron) is usually oriented in the direction of its speed. Once more we have broken symmetry.

But things get really exciting when we discover that the binding energy of the levorotatory molecules is slightly greater than the energy of the dextrorotatory (rightward) molecules. True, this difference is very tiny— on the order of one billionth of a billionth of an electron volt. But, under favorable conditions, it could generate the formation of chains of levorotatory molecules! The broken symmetry of beta radioactivity is, as we have seen, due to the properties of weak interactions. And they explain the tiny differences in the binding energy of the atoms and molecules. Isn't it conceivable that the difference in binding energies between the levorotatory and dextrorotatory molecules due to weak interactions is the cause of the asymmetry in the molecules that go to make up human beings? Here is a line of research which seems to me promising to say the least, connecting the appearance of life to the structure of the cosmos.

DISTRIBUTION AND FORMATION OF GALAXIES

We should now look in a different direction to discover new research perspectives. Before the time when the galaxies were touching, as we have said, they were not individualized and there was only a mass of gas from which they were formed. But we must admit that at the present time we have no satisfactory description of the mechanism of galaxy formation. What is certain, however, is that galaxies are apparently not distributed randomly through space and this might be one way of approaching this formidable question.

It has long been known that they appear to be distributed along "strings," "threads," or "chains," if you will. And it had been noted that the clusters of galaxies essentially occur at the intersections of the chains in question. But data analyses performed about five years ago showed that this distribution is not at all uniform. It was realized that the galaxies in our neighborhood, up to about a billion light-years away, were mostly distributed along the walls of huge "bubbles" (about 30 million light-years in diameter) which are not completely empty, but nearly so.

The word "bubble" is evocative, and for once the image is not deceptive. When we blow through a pipe into soapy water, for example, we form a cluster of bubbles and we can see that the soap film is distributed over surfaces whose intersections are lines, which themselves intersect at specific points. On quite another scale, of course, the galaxies appear to be distributed in the Universe on the surfaces of neighboring bubbles which cut into each other and intersect in a complex fashion.

Of course, we do not speak of "bubbles" solely for the sake of imagery in the trivial sense that we have just described. We do so to indicate, in the geometric sense, that the surfaces on which the galaxies are placed are closed and regular. To be more precise, we would have to say that they are distributed in space in such a way that an infinitely thin geometric surface could pass through them.

This finding is a very striking one. The diagrams furnished by Lapparent, the author of these studies, are very eloquent: they literally show the galaxies distributed in this way. If, for example, there are three bubbles in contact, when they are taken two by two there are three lines of intersection with one point in common. The galaxies and clusters of galaxies "appear" on these three lines and at this point. We should add that this is not just speculation. The locations of several hundred galaxies from a good-quality sampling have been plotted on these graphs just as on a map. One can "see" a number of "holes"—the "bubbles"—with edges rich in galaxies. But we have to understand what we mean by the word "see." Here it is used in the sense in which, if we are looking at a building head-on, we see the window sill and the inside of the window. We see the sash bars between panes and then these "holes" which correspond to panes that we cannot actually see.

Today, hardly anyone is contesting the validity of these diagrams, which for a time were suspected of arising from inadequacies and defects in observational data. The temptation immediately arises to interpret this regrouping as having been linked from the very beginning directly to the manner in which the galaxies were formed. In fact, it appears unlikely that these unexpected spatial distributions are closely dependent on the process of galaxy formation.

As I stated above, there is still no satisfactory description of this process. Perhaps the hypothetical cosmic "superstrings" were actually the agents of galaxy formation and the reason for their spatial distribution. But without adopting such a radical hypothesis, based on a kind of physics which is still being worked out, we can follow the very suggestive process devised by the Soviet physicist Y. Zeldovich about thirty years ago, even if his description remains essentially qualitative.

It can be illustrated by using a relatively simple optical analogy. If we consider a spherical mirror and a light source illuminating it, the light rays bathe a surface which is called a "caustic." It is interesting to note how the reflection of a light beam is distributed on a sheet of paper. As soon as the beams have been very slightly reflected outside the axis, they no longer go through the focus and are distributed by enveloping the "caustic."

Zeldovich's idea is stated as follows: if we are dealing with a medium in which disorderly movements are produced and we follow the trajectories of each of the participating particles, these trajectories will be tangential to surfaces analogous to the caustics and matter will assemble there to form galaxies. Hence we are dealing with a kind of simple production dynamic, not only of galaxies but of the places where they are formed. And in fact we can conceive that galaxies form where matter gathers, hence on surfaces and intersections of surfaces of the "caustic" type. However, we must acknowledge that, although Zeldovich's point of departure is very elegant, we are not talking about a physical explanation; it is rather a kind of metaphorical geometrization of an assumed process which ignores a formidable question linked to condensation, namely the elimination of heat.

This leads to other scenarios which attempt to bring this given into the picture. The best-known hypothesis is probably that of the cosmic superstrings, a model which responds to the earliest evidence of the distribution of the galaxies in strings; in fact, as we have just seen, the string image is not sufficient: we have to invoke the surfaces of bubbles. But the value of this model is the suggestion that there could have been gathering points of matter, attractive places for condensation, which could have been at the origin of the characteristic chain formation of the galaxies.

This model is linked to the efforts of physicists toward the unified field theory which calls on multidimensional geometries, some of which are believed to have a very limited range and underlie our four-dimensional Universe. It is within the framework of this complicated geometry that the lines of condensation of masses are said to have occurred. But, apart from the fact that this model does not take into account the now-established fact that galaxies are arranged on surfaces, it raises a problem which for the time being is insoluble: how could matter have become grouped at a time when the essential physical characteristics of the environment, when the Universe was made of a plasma of ionized gases, were very different from those that can be conceived today?

Since no experiments can be conducted on such an environment, the credibility of the model can rest only on its internal consistency. But even consistency is very difficult to achieve in this realm. The history of the last twenty years has proven that quite clearly.

Take only one example: if we consider the clusters of galaxies, namely the great systems that do not appear to have true structures, and if we consider a scale smaller than 15,000 light-years, what exactly are we studying? Correlations. One wonders what the probability is of finding the nearest galaxy to a cluster considered beyond a certain distance. So numerical studies are done on observational data to try to work out a law of correlations, and on this basis the attempt is made to construct a model satisfying this law. It is obvious from the studies that have been done that they involve only gravitational attraction: the effects of gravitation confer a distribution on the galaxies which is consistent with a law of correlation similar to that which has been observed. This is an analogous situation to what happens, for example, in the kinetic theory of gases, although with a different system of attraction. So we emerge with the feeling that this is a random phenomenon, describable in statistical terms.

It is true that when we get into the kinetic theory of gases we enter the realm of the random; but there are average values and we are dealing with so many objects that fluctuations become negligible and the predictability of these fluctuations turns out to be quite remarkable. But in astrophysics, when we deal with galaxies, the conditions are very different. Much as one would be tempted to search for simple causalities in the remotest times (which is what the partisans of cosmic superstrings do), later evolution reveals actions that appear to be completely local, not subject to global effects and escaping the effects of long-range interaction. Hence we can no longer simply reason in statistical terms.

HOW OLD IS THE UNIVERSE?

This being the case, it is easy to understand why the fascinating question of the "age" of the Universe is not settled, either. It should even be said that the question is often very poorly stated and leads the mind down paths which, from the scientific standpoint, are dead ends.

If we want to formulate this question correctly, the first step is to establish an order of magnitude. Given that "Hubble's constant" is a speed reckoned in millions of light-years, it is easy to see that a velocity divided by millions of hours corresponds to the converse of a time scale. Hence the converse of Hubble's constant allows us more or less to define the age of the Universe. We know that Hubble's constant is on the order of 25 km (16 mi) per second per million light-years. From this we can deduce the age of the Universe as being about 10 billion years. But it must be added that this figure reflects only a time scale and not a duration as the expression "age of the Universe" tends to suggest. The duration itself actually depends on the model by which movement is described.

In the simplest case, such as Friedmann's model, these 10 billion years correspond to an age of approximately 6.5 billion years. But estimates can be made based on other sources. For example, we could refer to the age of the oldest star systems in the vicinity of our galaxy. On this basis, we would obtain an age of between 12 and 18 billion years.

There are still other methods of dating. Some people use the radioactive elements we find on Earth and the products of radioactive decay. These very complicated techniques are based on the fact that each radioactive element has a known characteristic disintegration time called "radioactive decay," which can be measured in the laboratory. By very clever methods which I cannot describe here because of their sophistication, we arrive at an interval of between 12 and 18 billion years.

The situation merits examination with reference to the age of 6.5 billion years. On this scale, the background radiation of the sky was produced at a time when the Universe was 300,000 years old. Formation of the elements, then, occurred when the Universe was 100 seconds old. What we do know, but by simple reasoning without observational data, is that the time after which the quantum phenomena of relativity became negligible was very long ago and, in our scenario, corresponds to an age of 10^{-43} s. We must not forget that our physics cannot go beyond Planck's era.

Now we must reconcile the 6 billion years with the 12 to 18 billion. In trying to do this, we encounter some very major difficulties. One solution, for example, is to take the Lemaître model and postulate that expansion occurred but was slowed down by the effects of the cosmological constant, then resumed and continued until the present day. In general, if we allow for the existence of a cosmological constant, we are no longer tied to Friedmann's model of the Universe; we can accept a finite model of the Universe (spherical in a four-dimensional space), a close relative of Einstein's Universe of 1917, albeit expanding. So, with one additional parameter, we can solve quite a number of difficulties.

Whatever the validity of these scenarios, the important thing to keep in mind is that when astrophysicists are called upon to date the events which mark the history of the Universe, and consequently choose an origin of time, they are in no way establishing an origin for the Universe. Working backwards from the present day, they are constructing temporal points of reference that conform to the laws of physics available to them, which is quite a different thing. When they do this, they come up against a determinate time which, as we have seen, is approximately 10^{-43} s, and are totally incapable of saying what happened "before." The very question of "before" actually loses its meaning, because we are then entering a realm where gravitation (as general relativity leads us to believe) was most certainly of a "quantum" nature. But, at the present time we have no quantum theory of gravitation despite all the efforts made to unify general relativity and quantum mechanics. So we are entering an area that available scientific resources do not allow us to describe: about all that we can say about it is that it was a time when quantum fluctuations were becoming increasingly greater. Of course this is not very much if we continue to ask the haunting question of the origin of the Universe. But could the question itself be misleading?

III

INSTRUMENTS

AND REFLECTION

TELESCOPES, COMPUTERS, SATELLITES

The fact that astrophysics made the lightning progress we have just traced is due to ever closer cooperation on theoretical foundations worked out between several lines of research and even several disciplines which only recently became aware of each other. We have seen how the general as well as the special theory of relativity, quantum mechanics, nuclear physics, and finally particle physics have supplied not only results for analysis and instruments for use, but also hypotheses, some of which have proved extremely fertile while others, as I have emphasized, remain speculative.

But examination of the subject of our expanding Universe shows perhaps better than anything else how astrophysics remains an observational discipline. In fact, the great cosmological discoveries have been the result of systematic growth in measuring and observational capabilities, linked to technological progress. While one must be careful not to confuse scientific progress with technological progress as, unfortunately, we too often do; while we must not blame science for the crimes charged, sometimes not without reason, to technology—

or rather to the social and economic conditions under which technological process is exploited—we must nonetheless gauge how deeply technology has been involved in the progress of basic research.

Let us recall for one last time the decisive moment in history that we have just traced with broad brush strokes: the discovery of the receding galaxies by Hubble in 1926. First, the "redshift" of the galaxies had to be observed, and this would not have been possible without the building of the Mount Wilson telescope which, with its 2.5-m [100-in.] diameter, represented a long step forward in the power of our observational instruments; recording and interpreting this shift also required invention of the photographic plate and the spectrograph, the key instrument for spectral line analysis in the field of spectroscopy founded by Gustav Robert Kirchhoff in 1859.

Turning to the blackbody background radiation of the sky, I have already pointed out how much it owed to the testing of a new instrument by Bell Telephone Laboratories in 1963, even though, as I also said, the existence of such radiation had been postulated well before in purely theoretical ways.

This role of instruments has never ceased to grow in importance since that time. The increase in telescope diameter has allowed considerable progress to be made in the reception and study of infrared and radio waves. Giant telescopes are now available or soon will be: the telescope of the Teck Foundation in California, 12 m [39 ft] in diameter; the National New Technology Telescope in Hawaii,

16 m [52 ft] in diameter; and the similarly sized Very Large Telescope in Europe. These improvements which provide or promise an unprecedented accumulation of new data would not have been possible without the new technology which allows very large mirrors to be built that are light in weight and highly sensitive to infrared.

It is in fact well known that, starting in the 1960s when the first satellites were launched, observation could be free of the limiting effects of the Earth's atmosphere. The most spectacular discoveries were made in the x-ray and gamma ray ranges. First it was discovered that there were compact x-ray sources, consisting of double stars, namely two stars circling each other and revolving around their common center of gravity. In such cases, one of the two stars is extremely dense. Another type of x-ray source, still poorly understood today, was discovered in 1979 by the HEAO II satellite equipped with the Einstein telescope, which has a mirror capable of collecting x-rays and forming an x-ray picture of the sky in exactly the same way as an optical telescope. Through this instrument we discovered that, in the galaxies fairly near our own, what had thus far been treated as diffuse x-ray sources, in which x-rays were emitted over extensive areas, proved to be an agglomerate of compact sources.

In the field of gamma rays, several important discoveries were made in this way. In particular it was noticed that gamma radiation included flashes that could come in bursts. These are accounted for today by the sudden release of nuclear energy in matter captured at the surfaces of ultradense stars of very small radius, the famous "neutron stars." These stars, we must remember, were still relegated to the realm of fiction until 1967 at the time "pulsars" were discovered—stars which spin very rapidly on their own axes, once around in a second and sometimes much faster. In addition, gamma rays were detected coming from our own galaxy: it thus became possible to discover its structure and its production of cosmic rays. Yet another discovery was that gamma radiation coming from the background of the sky could be measured; its origin is not yet known but it appears to come from very remote regions of the Universe. Finally, in the central regions of our galaxy, emission of a type of radioactive aluminum produced in supernovas and disintegrating into magnesium-26 has been discovered.

Today, thanks to progress already achieved in telescopes, astrophysicists can "see" heavenly bodies that are over 10 billion light-years away! Soon they will be able to observe, if not galaxies in formation, at least galaxies that have newly formed. The ultimate proof of our expanding Universe would then come from observations of morphological and physical differences between galaxies as a function of their age.

Actually, if we were merely to mention progress in telescopy, we would be citing only one of the technological advances that has transformed astrophysics. To this we must add progress in imaging: the photographic plate available since the end of the last century has now been supplemented by new, far more sensitive photoelectric receivers. All of a sudden, mapping of the deep sky can make substantial progress by observing weaker objects.

Another recent and well-known technological innovation which has perhaps caused a more radical upheaval in the work of astrophysicists than in other disciplines was the appearance of the supercomputers. To take only one example: imagine the difficulties involved in studying the movements of the stars. In a double system, two stars revolve around the center of gravity of their own system, and there is a direct relationship between their masses, the distance between them, and the speed of their orbital movement. But this simple relationship vanishes as soon as there are more than two stars. The movements of several stars, although each is obeying an elementary law, are extremely complicated, with each star subject to the influence of all the others. So, only supercomputers have been able to calculate the movements of stars in such a system. At the present time we can create what is known as "digital simulation" of a system of several thousand stars: the movement of each one is "tracked" and, by graphic representation, we "see" this system being structured!

Finally, we have seen how the great particle accelerators built by physicists, such as the one at the CERN in Geneva, have provided astrophysicists with questions and confirmations, benefitting from cooperation that has now become institutionalized.

Admirable as the technological exploits played out on the stage of astrophysicists may be, I would venture to say that the deep human value of astrophysics is of quite another order. It seems to me to stem from its thought processes and the results it has achieved in our ambition to think the Universe through.

I have already described this thought process at sufficient length for me to refrain from dwelling on it further: to all those who make a point of denigrating science today, we must say that its path has been one of uninterrupted progress and constant deepening of our knowledge of reality. But I have also shown that this progress was not linear, as has often been erroneously pictured. It does not proceed by a simple accumulation of results, on theoretical foundations prepared in advance and using as its instrument a so-called received method. On the contrary, it is the fruit of ongoing requestioning and revision of established certainties and tested methods; a rethinking which has much to do with chance and which does not exclude, on the part of the researcher, the free association of ideas that appear to have little to do with one another, daring flights of speculation, trial and error, failures, second thoughts.... Indeed, we have seen that this progress is not in the least continuous: it proceeds by fits and starts. Tomorrow, a discovery may suddenly and quite unexpectedly reshuffle all the cards, even if the new truths are linked to those they have come to replace, assigning to them a new place in the edifice of science. It is no doubt this "revolutionary" aspect of scientific thought that makes it so suspect in the eyes of all the authorities charged with intellectual and social order and preservation. The history of investigations into the structure of the Universe and its genesis is replete with well-known episodes in which these authorities have mobilized against a new truth which upset ancient dogmas. If we can say that all science is "heretical" in essence, the study of the Universe must, by necessity, be more heretical than any other.

THE UNITY OF THE UNIVERSE

In my view, the most remarkable thing about the provisional results obtained from several thousand years of this study is that they show, and continue to confirm, the unity of the Universe.

The extraordinary enthusiasm which greeted the discovery of Neptune by Johann Gottfried Galle on the basis of a pure calculation made by Urbain Le Verrier from the Newtonian law of universal gravitation extended far beyond the scientific world. It was not unjustified: the law of universal gravitation which, since the publication of the *Principia Mathematica Philosophiae Naturalis* (1695) has guided the work of scientists and fascinated philosophers, took on a new significance. Not only did it explain the movements of known planets, but the discovery of a new planet conferred on it the unimpeachable status of representing reality. Moreover, it supplied a proof of that unity of the Universe on which Newton, after Galileo and Kepler, had focused his thought. And when, in 1915, Einstein formulated his new theory of gravitation, by which the slight anomaly already identified by Le Verrier in the orbit of Mercury could be accounted for, it was seen that, while it differed very little from Newtonian gravitation on the scale of the solar system, Einsteinian gravitation was important in the extreme cases of very dense bodies, bodies in rapid motion, or phenomena on the scale of the Universe. It is even more universal than Newtonian law whose validity it essentially safeguards on its own scale; a new step had been taken which confirmed the fascinating idea of the unity of the Universe.

The research whose successes and difficulties we have recounted confers extraordinary breadth and power on this idea.

In 1859, Kirchhoff discovered not only that every chemical element is characterized by the spectral lines it can emit—each line can be characterized as an emission of a very pure, almost monochromatic color—but that each element absorbs the lines it is able to emit. This discovery, made possible by contemporary progress in optics, immediately had far-reaching astronomical consequences because Father Secchi, observing some bright stars with a spectroscope, had identified certain major chemical elements such as hydrogen, sodium, and calcium. Thus, the idea of the unity of chemical composition of the cosmos was born. The same microscopic processes responsible on Earth for the process of radiation were found at work in the Sun and the stars. So far, everything has confirmed this idea: the production of spectral lines is governed by laws which, as we have seen, involve microscopic physical constants, the charge and the mass of the electron and the proton as well as the essential size of the subatomic world and Planck's constant. The presence of these same lines in the remotest stars we know shows that these physical constants have the same value everywhere in the Universe. The same physical mechanisms are generating the same spectra everywhere!

Another discovery already mentioned: the galaxies are strikingly similar in shape. This discovery of morphological similarity reinforces and completes the idea drawn from the identity of the chemical elements present in the Universe. It suggests that the same kind of physics governs the appearance of these shapes, and we can now see that gravitation determines the shapes of these great rotating systems, the formation of spiral arms being due to the instabilities caused by rotation and the tidal effects found when one galaxy passes near another being responsible for complex characteristic shapes.

We have also seen how gravitational forces govern the movement of the planets, the stars, and the galaxies, and we know that electromagnetic forces fix the structure of atoms and the production of light. But I have stated that two other forces govern the fine structures of matter: the force which binds together protons and neutrons in the nucleus of the atom, called the strong interaction; and the force involved in radioactive processes which binds neutrinos and antineutrinos, those particles virtually without mass, and is called the weak interaction.

These interactions have been demonstrated in the laboratory. But, little by little, they have been discovered as playing a role in the hearts of stars and at the very confines of the Universe! It is strong interactions that account for the energy of the stars; weak interactions are found at the center of the Sun, where they are responsible for producing neutrinos. Finally, as we now know, it is the joint effect of strong and weak interactions which is at the root of the primeval production of light elements (hydrogen and helium) found everywhere in the Universe.

How can we fail to be dazzled by this prodigious unity of the physical world? How can we not subscribe to that grandiose ambition of physicists who seek to unify the four forces of nature?

Are we forced to admit that this unity "disenchants" the world because of excessive uniformity? Not at all, because it is not in the least incompatible with the fantastic diversity of shapes and objects which make up the Universe. With regard to astronomical objects alone, this diversity already defies the imagination. The unity of physical laws does not prevent, through differences in parameters (era of formation, mass of stars, rotation, and multiplicity) the manifestation of diversity to the point where we can say that there is no such thing as two identical galaxies, and even two identical stars. Especially since we know that there are 100 billion stars in our galaxy alone!

BASIC RESEARCH

Important progress in stellar and galactic astrophysics has been made, as we have just seen, through measurements made in space in the infrared, x-ray, and gamma ray ranges. Now, it is well known that the development of increasingly powerful launchers, placing heavier and heavier and more and more sophisticated satellites in space, is linked to military needs, whether we are talking about ground surveillance by spy satellites or the preparation of new modalities of deterrence or confrontation, as shown by the preliminary studies in the Strategic Defense Initiative (SDI or "Star Wars").

Must we conclude that the development of astrophysics will henceforth be so bound up with military objectives that its scientific objectivity must be called into question? Because the space programs that allow us to advance our knowledge are largely underwritten by military budgets, must this discipline be viewed as an auxiliary, if not a handmaiden of the martial arts and therefore, considering present-day conditions, a threat to civilization? And can we really speak of the "militarization of research" and accuse astrophysicists of being naive or cynical accomplices of programs controlled by the most unscientific reasoning that jeopardize our entire planet?

These questions have frequently been raised and these criticisms and attacks have been made, often with extreme virulence, over the last twenty years or so. They would seem to me based on a series of very profound misunderstandings.

The first has to do with the role of instruments in research. It is perfectly true that many of them were developed and rapidly improved to meet military requirements. Apart from satellites, we can readily think of radar during World War II, and progress in photometry. But are these instruments "part and parcel" of astrophysics, to the point that it should be imported and then stamped with the imprint of their initial objectives? Quite obviously not. They admit of many applications, and physics and astronomy are simply two of the beneficiaries, in terms of their own theoretical concerns.

Because the Concorde could be used to observe a solar eclipse, would we call this supersonic aircraft an astrophysical instrument? And would we draw the same kind of conclusion about the relationship between this discipline and the aeronautical industry? This would be so manifestly absurd that no one would ever dream of it. Yet, in essence, the case is a similar one.

Generally, progress in astrophysical observation benefits from technological developments which have their own history and logic. The use of these instruments for astronomy has been and remains marginal. Just think of the artificial satellites and the time it took after the launching of the first Sputnik in 1957 for observational instruments to be placed on board satellites!

As for the logic of using these instruments, far from being dictated or even inflected, not to say infected, by the use of the programs that motivated their development, it has never been anything but research into astrophysics itself. The choice of subjects, the decision to make this or that observation to test this or that hypothesis, derives exclusively from this research and its imperatives.

The second misunderstanding is based on a pernicious confusion between research and the social applications of its results—industrial, military, and other applications. The questions that these applications raise are very serious ones. But they can be solved only by seeing them for what they are: not scientific questions but political questions. It is a grave mistake to imagine that these applications, whatever they may be, inevitably and fatally are a result of fundamental research itself, in which they lie latent, so to speak. This would be to deny the timing and reasons for choice, hence the responsibility, of the decision-maker or decision-makers; it would at the same time ignore the weightiest question that arises for democracies today in these matters: how are citizens to be given the tools for understanding all the ins and outs of the decisions that are to be made? How can they be brought into the decision-making process and participate actively through public scrutiny and debate of the choice?

In the same way, when we look at funding, is it not obvious that research requiring huge investment outlays depends on budget allocations and hence on the authorities that allocate the funds? Of course, and this too is a serious question. It needs to be asked nonetheless. This is not a moral question; it is not because an area of research is funded by, shall we say, the military that research itself is "contaminated." On the other hand, there is a legitimate fear that the major programs will soak up all the funds and that free research by small groups—the research that has often been the most fertile in physics—would be sacrificed. But how can this question be examined with the seriousness it deserves if "science" is overwhelmed by critics who identify it purely and simply with the technological consequences of its major programs?

The reader will understand why, in a recent book, La Science Menacée [Science threatened] (Odile Jacob, 1989), I became alarmed by the attitude toward science of a large number of intellectuals and many politicians. Yes, science is threatened by their ignorance of its reality, by the confusion maintained about the goals and approaches of basic research, by the profits reaped by certain powerful economic, financial, and ideological groups that exploit this ignorance and confusion. To round out this catastrophic view, we must include the sort of confusion that exists between the knowledge of scientists and their actions as citizens. Although they have greater civic responsibilities than others, what is nonetheless at stake is a political matter in which the individual and the governments using him or her are engaged. In the final analysis, what is at issue is not science, nor scientific knowledge, but the exercise of democratic control over decision-makers and decision-making systems. The matter is too grave for me not to expand on it a little by accentuating the contrast between what is said about science and the reality of the intellectual conquests I have just outlined.

In the broadest and noblest sense of the term, the problem is a political one: it involves education, culture, and decision-making mechanisms in our societies.

First of all, how can we restrain our indignation at the way science is taught? The leaps of the mind, the burning desire for knowledge that drives scientific research, of which astrophysics is a singularly uplifting illustration, is precisely what has been repressed and stifled, to the point of being irretrievably lost to our students. Today's education gives them no way of acceding to the burning issues of science today that we have just described. And this is just as true at the university level as for education at other levels.

The damage is already done in high school: instead of enabling students to understand the nature of the scientific approach, instead of resconstructing the history of the issues with which minds grappled, students are given formulas to learn and recite. These formulas appear to have dropped from the sky; they are presented and (usually) taught dogmatically. And then, by the tried and true scholastic method, the "merit" of a student is judged merely by testing his or her memory! Served up this way, science can no longer appear as an exercise of the mind, a demanding and daring one; it becomes an instrument of intellectual submission. That such submission leads to academic success solves nothing—on the contrary. We can understand why teachers of the physical sciences complain about the inadequacy of the resources made available to high schools to teach the nature of experimental science and hence to show what a formula actually means.

How can we fail to be astonished that scientific teaching is altogether designed and implemented, regardless of the audience to which it is addressed, to lead to the highest academic diplomas? Manual after manual, year after year, the formulas are strung out in a process of accumulation whose end is the total sum of formulas a highly qualified researcher is supposed to know. But how many pupils and students want to become such a researcher when they grow up? And what is one to say of the teaching of science in the literary curriculum where students are merely served up bite-size pieces of this dogma?

Do things improve in higher education? Certainly not, but for other reasons, because the French university system, being historically founded on the requirements for teaching, has in the last twenty years suffered a recruiting breakdown which puts its future at stake. The efforts of professors to do research are compromised by excessive administrative duties.

These remarks, which apply to France, are not, I think, without pertinence in many countries, even those such as the United States that others are inclined to take as models.

If we add to this deficiency the role played by the media in presenting and transmitting the culture of our times, the picture grows even darker, because when it comes to science the media stress one note and one note alone: the spectacular. They need events, which they are quite capable of creating out of whole cloth, and events with an emotive charge. This is what leads them alternatively to "diabolize" science or "deify" its results quite uncritically. Astrophysics and cosmology are, from this standpoint, particularly treasured because the issues with which they are involved speak to the imagination.

To mention only television, who could claim that the numerous programs devoted to these subjects have really contributed in the past twenty years to conveying to the public at large the true content and direction of scientific research?

The nadir is reached when a work hits the headlines only because of the well-publicized disability of its author and the fact that on several occasions he or she inserts the word "God" in the middle of the equations the author used!

Now that education and culture have done their work, it is scarcely surprising that some of the finest minds share with the masses a concept of science which brings basic research to an impasse. To them, such research is purely and simply equated with technology—an object of veneration for some, who expect from it the solutions to all social or even psychological problems; an object of fear for others who attribute to it all the tragedies that now weigh down our society: the military use of nuclear energy, the ecological devastation of the planet, the impoverishment of the Third World, and so on.

In this situation, the politicians—decision-makers who suffer from the same prejudices as other people—calculate what is in their interests. Since these calculations translate immediately into budget allocations, the matter becomes quite serious because the choices made can weigh very heavily on the future of research.

Two features of basic research that are particularly manifest in the case of astrophysics should be borne in mind: since its goal is the advancement of knowledge, research operates on a long-term basis and cannot be sure of its results until it has obtained them, sometimes by accident; hence it involves an irreducible element of chance. Today it demands huge investments: a supertelescope, not to mention a particle accelerator or a space expedition, can easily cost a couple of hundred million dollars.

When choices have to be made, politicians inevitably tend to reason not in the long term, as they should in such matters, but in the short or medium term; also they tend to favor research which they think will bring some tangible advantage, whether this advantage is economic, military, or simply one of prestige. To cite only one example which affects our field of research, take the famous Strategic Defense Initiative approved by Ronald Reagan in 1983. I have already had occasion to demonstrate the eminently pernicious nature of this decision. In particular, I analyzed in detail the technical difficulties of an ideologically mystifying and scientifically poorly controlled project; I must emphasize here that the total cost of the project is estimated to be ten thousand billion dollars! The conclusion is inescapable that it will affect the direction of research and, by a well-known cascade phenomenon, the recruiting and training of researchers. It is said that such a project will "drive" basic research by unexpected funding of some of its areas. But this is pure illusion, as has been shown by Hans Bethe, Garwin, Carl Sagan, and Victor Weisskopf, who originated the oath signed by thousands of American scientists vowing never to work for SDI.

THE RETURN OF GOD?

Thus misunderstood, basic research, which is at the very heart of science, becomes eminently vulnerable. Its results may be drawn into the first ideology to come along. We see this, to return to this question one last time, with the expansion of the Universe.

Whatever they may say, this idea, as it has been patiently built up by astrophysicists to account for the observational data they have been amassing since the beginning of the century, does not resuscitate the old, ancient notion of origin; today it does not even speak of a beginning of the Universe. So why succumb, as so many reputable thinkers and even astrophysicists have done, to the age-old cosmogonic temptation? Hastening to pose the question of origin and describe a new genesis is to let the mind fall into a trap, engaging it in the same line of argument in which it has been bogged down for thousands of years; it is fleeing from a difficulty which may very well be naturally inherent in the human mind, that of thinking in terms of the infinite. To insist on assigning an origin to the Universe at any price is to desire to put an end at any price to the tortures that the infinite has never ceased to inflict on us. We know that the greatest thoughts are those that have been willing to confront it.

It may be true that the notion of origin, and its anguished implications for the destiny of the human race, are philosophically legitimate, although there has been no lack of philosophers, the greatest among them those who have challenged the terms and denounced the downhill path of religion as a path to intellectual submission. But the deception begins when astrophysics is called to the rescue, when it is claimed that this or that latest interpretation of experimental data finally allows us to put our finger on an origin as being scientifically established; more serious still, when what is only interpretation or conjecture is announced as "fact."

All that we are entitled to say about the past of the Universe, I have tried to express simply in this short book. We have, I repeat, observed residual radiation from a time when the Universe was homogeneous and very hot. In this burning Universe, reactions produced extreme temperatures. Our comprehension of the Universe goes back to the time when it cooled down to 100,000 billion billion degrees Kelvin. And if we want to "translate" this temperature into chronological terms, we would say that this is a fraction of 10^{-43} s after what we establish as the beginning of this process, based on present knowledge.

What happened before? We have to admit that no one knows. Everyone talks of the big bang and thinks of it as a "historic" event that science "discovered." And this supposed event is described as a gigantic explosion that flung material into space. But this is mere talk and says nothing because the big bang, if big bang there was, occurred at a moment when there was neither matter nor space as understood by modern physics. Indeed this is why modern physics has nothing to say about this big bang; this is why it is only a hypothesis. All that can be said is that this hypothesis is for the time being, if not imposed, at least strongly suggested by the status of knowledge to account for the point where it stops.

As we have also seen, daring and ingenious models have been devised to account for the formation and distribution of the galaxies. One of them speaks of extraordinary particles that united to form gravitational sites of attraction. There has also been talk of superstrings. But it should be noted that both phenomena could only be pinpointed at 10^{-35} s, and should hence not be allowed to confuse matters. What is more, we must never forget that these are not established "facts," but invented scenarios that cannot be confirmed by observation. They are scenarios for apprehending the Universe. And for the future of research it is at the very least prudent not to take hypothesis or scenario for fact if we want to retain alternative options. We know from experience how thinking can be stymied by such confusion.

Unfortunately, in the case of a process which touches the boundaries of the Universe, many forces, some of them irrational, are brought to weigh heavy on the scientist's thinking and to interfere with it. We thus witness the resurgence, under the pompous name of "anthropic principle," of finalistic or teleological arguments which amount to attributing the order of the Universe to a divine purpose.

The core of the argument rests on the analysis of four forces of nature. It appears that if the fundamental constants that determine the value of these forces had been only very slightly different, the appearance of life, and ultimately of the intelligent human being, would have been impossible. For example, if the strong interactions had been less intense, combustion of hydrogen could not have occurred in the stars and only gravitational contraction would have been the source of short-lived radiation; the production of heavy elements would have been impossible. Likewise, a stronger gravitational constant would have led to a mode of evolution of the stars incompatible with the conditions for the emergence of life. The "anthropic principle," announced in 1974 by B. Carter of the Meudon Observatory, states that the fundamental constants have the value they have because they made possible the appearance of the human species. F. Dyson, of the famous Institute for Advanced Study at Princeton, is even more radical when he writes that "the remarkable harmony between the structure of the Universe and the needs of life and intelligence is a manifestation of the importance of the mind on the agenda of time."

This type of discourse which leads everyone to God is today common currency; divine "creation" is said to have consisted in choosing the fundamental constants in order to trigger the appearance of human beings!

These arguments are as eternal as the anguish inherent in the human condition; they hark back to what we might have believed to be a bygone era, when religion sought evidence in the order of the world to reinforce the authority of the Scriptures. The fact that eminent scientists adopt Holy Writ as their own and write about it does not make them less religious, less outward-looking, less averse to the effort of scientific thinking.

Let us hope that the churches, having learned from the Galileo affair and then from that of Darwin, will understand that in fact faith has more to fear than to hope for when it gets mixed up with science. As for scientists, although some of them may for personal reasons have a mystical bent, the great majority know where to draw the line. Finally, we may hope for something that is perhaps the most difficult of all today, namely that researchers will be able to resist the powerful call of the media urging them to enter the game of mystification, where science has everything to lose and which, moreover, attests to a profound contempt for the citizen by mortgaging the most precious human possession, the freedom to think.

Certainly citizens have the right to dream; in fact, for me one of the greatest attractions of astrophysics is that it ceaselessly renews the dream and sends it along rocky paths; but they are also entitled to know, and to know first of all, where knowledge stops and dreaming begins!

BIBLIOGRAPHY

AUDOUZE, J., *Aujourd'hui l'Univers* [The Universe today*], Belfond, 1989.

COHEN-TANNOUDJI, G. and SPIRO, M., *La Matière espace-temps* [The space-time material*], Fayard, 1986.

HOYLÉ, F., *Astronomy and Cosmology. A Modern Course*. W.H. Freeman & Co., San Francisco, 1975.

KOYRÉ, A. *From the Closed World to the Infinite Universe*. Hopkins Press, Baltimore, 1968.

PECKER, J.-C., series under the editorial supervision of, *Astronomie Flammarion*, Flammarion, 1985.

REEVES, H., *L'Heure de s'enivrer* [The intoxicating hour*], Le Seuil, 1986.

SCHATZMAN, E., *Les Enfants d'Uranie* [The children of Urania*], Le Seuil, 1986; *La Science menacée* [Science threatened*], Odile Jacob, 1989.

SCHNEIDER, J., series under the editorial supervision of, *Aux confins de l'Univers* [On the borders of the Universe*], Fayard-Fondaton Diderot, 1987.

WEINBERG, S., *Les Trois Premières Minutes de l'Univers* [The first three minutes of the Universe*], Le Seuil, 1979.

---

* These references have not been published in English.